Accelerated
Reader

D0742570

FIGHTING FORCES
ON LAND

M2A2 BRADLEY
FIGHTING
VEHICLE

DAVID BAKER

Rourke

Publishing LLC
Vero Beach, Florida 32964

www.rourkepublishing.com

PHOTO CREDITS: All photos courtesy United States Department of Defense, United States Department of the Army

Title page: *An M2 Bradley Infantry Fighting Vehicle of the 1st Cavalry Division on patrol in Iraq. Note the track-side armor plates for protection from shrapnel and small arms fire.*

Editor: Robert Stengard-Olliges

Library of Congress Cataloging-in-Publication Data

Baker, David, 1944-
  M2A2 Bradley fighting vehicle / David Baker.
    p. cm. -- (Fighting forces on land)
  Includes bibliographical references and index.
  ISBN 1-60044-249-8 (alk. paper)
  1. M2 Bradley infantry fighting vehicle--Juvenile literature. I. Title.
  II. Series. 9-17-07
  UG446.5.B2349 2007
  623.7'475--dc22
                                        2006011681

Printed in the USA

CG/CG

Rourke Publishing

www.rourkepublishing.com – sales@rourkepublishing.com
Post Office Box 3328, Vero Beach, FL 32964

TABLE OF CONTENTS

CHAPTER ONE

KEEPING UP WITH THE TANKS

When the M1A1 Abrams main battle tank deployed with US Army units in 1980, it came as a shock to the infantry to realize they did not have a fighting vehicle that could keep up with it in battle.

▲

The turret supports a 25 mm M242 dual-feed chain gun with automatic stabilization and 300 ready rounds with a further 600 stowed.

With sufficient speed to keep up with the Abrams main battle tank, the Bradley allows the infantry to support heavy armor and not fall behind on the battlefield.

The Bradley is capable of holding its own in a firefight but its main job is to get the infantry into remote places and provide cover with its weapons until the job is done.

This Bradley in Baghdad is doing some fast maneuvering to address fire coming from across the sidewalk, emphasizing the cooperation between driver and commander as the vehicle drives to position while the gunner lays his turret on the target.

Tanks and infantry frequently **cooperate** in battle, the former pushing through heavily defended areas clearing a path for armored personnel carriers and Infantry Fighting Vehicles, or IFVs. Bradley IFVs provide the muscle to allow soldiers to destroy small pockets of **resistance** as well as enemy tanks and armored fighting vehicles.

CHAPTER TWO

BUILDING TO FIGHT

As a combat type, the Infantry Fighting Vehicle first appeared when Soviet forces introduced the BMP-1during the late 1960s. Designed to support rapid maneuvers during an assault involving fast moving tanks and tracked vehicles, the BMP-1 has forward slanted armor in the front. Equipped with a smoothbore gun and anti-tank guided missiles, it was a **potent** threat. The US forces needed a powerful counter to this new form of combat anti-tank vehicle.

▲

The Bradley has six road wheels and two sprocket drives to each track with idler wheels under the armored side plates visible on this view of a Bradley backing away from a fight.

▲

Protected with armor plate, the Bradley is suited to a wide variety of terrain and climate. This M2 is supporting infantry in Afghanistan.

Combining a need to match the BMP and to support fast moving battlefield **operations** involving the Abrams main battle tank, the M2 Bradley emerged in 1981 partly as a replacement for the M113 Gavin but also to open a new type of infantry support role. A variant based on the same chassis, the M3 entered service as a Cavalry Fighting Vehicle for two-man **reconnaissance** missions. The M2 on the other hand would carry a crew of three and a six-man infantry squad.

▲

Stripped to the bar steel, this Bradley at speed displays the basic structure of the vehicle usually obscured with armor plating and the assorted stores and equipment that surrounds soldiers wherever they go.

▲

Rough terrain is no problem for the tracked Bradley but street performance is equally important when US forces are frequently engaged in urban warfare and attack can come from street side houses and building. This Bradley carries box-armor designed to detonate shells before they reach the steel structure of the tank and cause serious damage.

Hitting the Enemy

Although evolving from a Soviet challenge and doubling as a fully-fledged, hard-hitting, combat vehicle for the infantry, the Bradley would match its Russian counterpart by carrying missiles in addition to a gun. Named after World War II (1939-1945) General Omar Bradley, the M2 carries a 25 mm chain gun, TOW (Tube-launched, Optically-tracked, Wire command-link guided) missiles for knocking out tanks and armored vehicles and carries a 7.62 mm machine gun as secondary **armament**. The chain gun fires 200 rounds a minute and is a highly effective weapon to a distance of 1.5 miles but the twin TOW missiles can destroy any armored tank currently in service anywhere in the world to a distance of more than two miles.

▲

The wedge shaped revolving gun turret is shaped to deflect incoming artillery shells and small arms fire and has additional armor that can be added for extra protection from hits. This driver gets to feel fresh air by opening the generous lid hatch.

▲

Secure and closed up, the driver is well protected from incoming fire. Note the two M257 smoke grenade dischargers, each with four tubes, on either side of the 25 mm chain gun.

The two TOW missiles are carried in a collapsible, armored launch rack to the left of the chain gun and when fired reach a speed of almost Mach 1 – the speed of sound – at impact. The Bradley must stop to fire the missiles and an infantryman reloads the launch tubes from the back of the vehicle. A special **armor** plated hatch protects the infantryman from enemy fire.

▲
In firefights, a steel dome-like hatch protects the driver; forward and limited side-view vision blocks provide sight with the hatch closed.

Because the vehicle is built largely of aluminum it was thought by some that it would be easy to **ignite** – aluminum is notoriously flammable when heated as it would be if hit with certain types of tank shells designed to burn metal. In fact, the Bradley performs well and it has not succumbed to this type of ammunition on the battlefield.

▲

A detailed photo of the Bradley's side skirt armor and explosive reaction armor (ERA). Ninty-six ERA tiles are fixed to the sides, turret, and front of the Bradley.

CHAPTER FOUR

DESERT FIGHTING

To date there are more than 6,800 Bradleys of which 4,600 are the M2 Infantry Fighting Vehicle. They have been used in the Gulf during the military operation in 1991 to **evict** Iraq from Kuwait and again in 2003 to invade Iraq and **depose** Saddam Hussein.

▲

Night fighting is an essential part of the modern battlefield. Bradley's must fight in all conditions, day or night.

▲

Soldiers in their Bradley fighting vehicle from the 15th Infantry Regiment, 3rd Infantry Division, support Iraq troops in their search for insurgents near Samarra, Iraq.

During the 1991 Gulf War, the 2,200 Bradleys involved in Operation Desert Storm fought hard and heavy, destroying more Iraqi tanks than the M1A1 Abrams. During that conflict, twenty M2s and M3s were lost but only three were due to enemy action. Unfortunately, 17 were lost to friendly fire incidents when they were incorrectly identified as enemy vehicles. To avoid this in the future the US Army applied special infrared identification panels and other highly visible markings to the exterior.

▲

Night mail. Using low-light level goggles this infantryman walks a Bradley packed with cases.

▲

A battle scarred Bradley shows off its generous rear exit and entry door.

▲

Unlike most infantry and armored vehicles, the Bradley provides generous hatch and door openings and flat top surfaces make a great place to catch a snack.

During Operation Iraqi Freedom, beginning in 2003, the Bradley proved **vulnerable** to rocket propelled grenades and roadside bombs but the **doctrine** has been to abandon the vehicle and save the personnel, a policy that has saved lives at the expense of vehicle.

CHAPTER FIVE

UPGRADES

Like many military vehicles, the Bradley has been upgraded and adapted to carry out different missions. The basic M2 weighs 23 tons unloaded and carries an additional eight tons of soldiers and equipment. Its 600 hp diesel engine is more than twice as powerful as the M113 Gavin and it has a top road speed of 38 mph and a range of 250 miles between fill-ups.

▲

Under most conditions, the M2 is capable of speeds in excess of 35 mph and that gives this IFV valuable advantage for getting into and out of trouble fast and with maximum safety.

Not all duties are down and dirty. This M2 is returning from airport patrol in Iraq, one of many jobs this ubiquitous vehicle and its personnel perform.

▲

The rear door serves as an armored lid to protect infantrymen buttoned up inside. Bradleys work in packs where teamwork is essential. These IFV's also carry box-armor to predetonate incoming rounds of exploding ammunition.

Loaded with stores, supplies, equipment, weapons, and rolls of barbed wire, the Bradley can play the role of the ultimate pack mule and is always used in coordination with infantry and other ground troops. ▶

The M3A3, the latest version of the Bradley entering service in 2000, carries seven infantrymen as well as the three-man crew and has additional protection provided by titanium roof armor. All Bradleys built since 1988 have side protection against 30 mm projectiles.

Constant improvements ease crew workload by providing eye-safe laser ranging equipment, tactical direction and navigation through **GPS** satellites, and a digital compass more accurate than the instrument version. Servicing too is easier, with an electric lift for the engine access door, outside stowage for personal gear, and additional ammunition boxes inside.

CHAPTER SIX

DEVELOPMENTS

The Bradley has evolved from an Infantry Fighting Vehicle into a mission-customized base vehicle, some replacing TOW anti-tank missiles with the Stinger anti-aircraft missile for local air defense, others doubling as fire support vehicles directing artillery barrages from towed and self-propelled howitzers. There is even a version for army medical evacuation work.

▲
A Bradley with twin TOW anti-tank missiles maintains watch for Serbian forces in Kosovo as part of the United Nations operation.

The Bradley has carved out a special **niche** in battlefield operations while demonstrating, especially in Iraq, that it is the right sort of tracked vehicle for a mix of open desert and urban street fighting, supporting the infantryman with hard-hitting firepower destructive against any military vehicle yet built.

▲

The M2 has proved rugged and durable in hostile conditions and poor weather, essential characteristics for the fighting soldier who must rely on the IFV for transport and protection from gunfire.

Glossary

armament (AR muh muhnt) – weapons and other equipment used for fighting wars

armor (AR mur) – strong metal protection for military vehicles

cooperate (koh OP uh rate) – to work together

depose (de POZE) – to take away power or authority

doctrine (DOK trin) – a belief or teaching of a group of people

evict (i VIKT) – to force someone to move out

GPS (Global Positioning System) – a satellite navigation system that allows for accurate targeting of missiles and improved command and control of forces

ignite (ig NITE) – to catch fire

niche (NICH) – a position for which someone is especially suited

operations (op uh RAY shuhnz) – an event that has been carefully planned and involves a lot of people

potent (POHT uhnt) – powerful or strong

reconnaissance (re KUHN uh senss) – the active gathering of information about an enemy, or other conditions, by physical observation

resistance (ri ZISS tuhnss) – the ability to fight off or overcome something

vulnerable (VUHL nur uh buhl) – in a weak position and likely to be hurt of damaged in some way

INDEX

FURTHER READING

Hunnicutt, R.P. *Bradley: A History of American Fighting and Support Vehicles*. Presidio, 1999

WEBSITES TO VISIT

http://www.fas.org/man/dod-101/sys/land/m2.htm
http://www.battletanks.com/m2_Bradley.htm

ABOUT THE AUTHOR

David Baker is a specialist in defense and space programs, author of more than 60 books and consultant to many government and industry organizations. David is also a lecturer and policy analyst and regularly visits many countries around the world in the pursuit of his work.